Dalton Moore

Friendly Bacteria

Fight and prevent disease with healthy intestinal flora

alive books

Vancouver
Canada

Contents

All About Friendly Bacteria

Note: Conversions in this book (from imperial to metric) are not exact. They have been rounded to the nearest measurement for convenience. Exact measurements are given in imperial. The recipes in this book are by no means to be taken as therapeutic. They simply promote the philosophy of both the author and *alive* books in relation to whole foods, health and nutrition, while incorporating the practical advice given by the author in the first section of the book.

Healthy Bacteria-Boosting Recipes

All About
Friendly Bacteria

Friendly bacteria are welcome tenants in our bodies; they
are essential to maintaining a state of good health.

Introduction . . . .

I spent my formative years, as most people do, totally oblivious to the amazing world within my own body. I was led to believe that all bacteria were bad. The word "bacteria" was usually accompanied by the word "infection," and this was something to be taken care of before it caused severe complications that would surely require hospital intervention. If the infection was internal, an antibiotic was quickly administered. Little did I suspect that there were also good, friendly bacteria that were actually beneficial and even essential to my well being that were also being destroyed by the antibiotic medication that I had been given.

Although your initial reaction to bacteria might be negative, in fact, we couldn't live without them.

This book deals with the tiniest of living organisms, bacteria. These microscopic animals cover the planet: they are in our food, in the air, in the water; everywhere, including inside our bodies. Although your initial reaction to bacteria might be negative, in fact, we couldn't live without them. Treat them well and they can help you maintain good health without drugs, and they can ease a wide variety of all-too-common ailments, from acne to cancer. In relation to the human body, these little animals fall into two categories: the good guys and the bad guys.

Good bacteria do a wealth of good in the body. Most importantly, they keep potentially pathogenic (bad) bacteria from becoming established. Friendly bacteria can help to prevent disease and have the ability to manufacture vitamins and natural antibiotics. When good bacteria are killed by man-made antibiotics we are left with an imbalance of good and bad bacteria, which results in illness and disease.

Good bacteria are welcome tenants in our bodies: they are essential to maintaining a state of good health. These tiny microorganisms live in a cooperative, symbiotic relationship within each of us. Our friendly bacterial community consists of many different strains of bacteria, each with a critical job to do to maintain the good health of the host—you and I. These various families, in all their diversity, are with us from the day we are born and hopefully remain with us throughout our lives.

It took me years of research and personal contact with the experts in the complex world of friendly bacteria before I came to appreciate the crucial role that bacteria play in the day-to-day functioning of a healthy human body. Bacteria live everywhere in our bodies—in our eyes and mouths, on our skin, and in our gastrointestinal tract. I will concentrate on this last group in this book. It is my intention to lift the veil of mystery from the essential roles that good bacteria play within us. You will discover that everything is not simply black or white; there are many contributing factors in maintaining a healthy body.

Ensuring that you support the friendly bacteria within is only part of the picture. Diet and lifestyle are significant contributing factors, too, and proper day-to-day regularity is essential to good health. When our friendly bacterial colonies have been devastated by antibiotics, cortisone, junk food, and pollutants, it is important to know what to look for when choosing a bacterial supplement.

What are Bacteria Anyway? . . .

Everyone is aware of the term "antibiotic," but few can define "probiotic." The Greek word that describes life or living is *biotikos*, so "antibiotic" means something that destroys living organisms, while "probiotic" is some-

Corina Messerschmidt

thing that preserves living organisms. Bacteria are very tiny, living micro-organisms (animals) that have been around since the beginning of time. Bacteria are so tiny you could put one billion of them on the head of a pin and never know they were there. Bacteria are everywhere: in the air we breathe, the water we drink and in great abundance on the food we eat. Don't be alarmed: this is the way it was meant to be.

Friendly bacteria work tirelessly to create a Garden of Eden within each of us.

Bacteria are like other groups in the animal kingdom. If we compare them to dogs, for example, we can see that there are many different groups (dogs, wolves, foxes) and several strains, or breeds, of each (terriers, spaniels, retrievers). They are all dogs, but each is different and distinct. So it is with bacteria. There are many distinct families, with many differing individuals within each family. For instance, the lacto-bacilli are one genus; *acidophilus* and *bulgaricus* are species of that genus.

Good bacteria live very happily within our bodies by the trillions, especially in our gastrointestinal tract. They set up house-keeping in areas that are warm, moist and–most importantly–have a hospitable pH level. The ideal pH environment is slightly acidic, somewhere between pH3 and pH6. PH stands for Potential of Hydrogen and ranges from total acid (pH0) to total alkaline (pH14); neutral is pH7 (the pH value of water is 7). Acidity is crucial for two reasons: an acidic environment supports good bacteria, and actively discourages the bad ones.

These remarkable and complex micro-organisms play a vital role in our well being. If properly nurtured and supported, friendly bacteria work tirelessly to create a Garden of Eden within each of us. Friendly bacteria are responsible for the production of various vital nutrients, including vitamins B_2 (riboflavin), B_3 (niacin), B_5 (pantothenic acid), B_6 (pyridoxine) and B_{12} (cyanocobalamine); biotin; folic acid; vitamin K; lactase enzymes; and hydrogen peroxide. Their work also includes predigestion of proteins and manufacturing acidophilin (a natural antibiotic). Whew! And for all that, all they ask is a nice warm place to call home!

Friendly bacteria are especially active in the colon, where most of the manufacturing processes take place, toxins are dealt with and elimination of waste is facilitated. A healthy colon contains approximately 85 percent good bacteria and 15 percent putrefactive coliform bacteria (these are not good). If this ratio is maintained, the colon is very healthy, but unfortunately, with today's lifestyle and fast-food diets, some of us are lucky to have a 50:50 ratio.

> ### The Search for Information
> In the past, there was very little readable, easily understood information available on good bacteria. I was prompted to investigate and to promote a better understanding of friendly bacteria when I entered the field of natural healing. My background was in the area of pharmaceuticals, doctors, clinics and hospitals. Changing my career from one in the medical/pharmaceutical field to the study of herbal medicine and natural, alternative healing was challenging to say the least. My mentor in my new career was Siegfried Gursche (the father of natural health in Canada and publisher of *alive* books), who introduced me to a world that we all take for granted. My experience was with prescription drugs, in treating just the symptoms of an ailment, so this natural approach was totally new to me.
>
> I was eager to explore the complex and little-understood role that friendly bacteria (micro-organisms) play in maintaining good health. But there was simply no reference material readily available in Canada. The closest I came was a book titled *Gastro Intestinal Physiology* by Drs. D.N. Grainger and J.A. Barroman. Although very technical, this book whetted my appetite. Adding to my desire to learn all I could about friendly bacteria was Siegfried's decision to add a bacterial supplement to his company's line of natural alternative supplemental and herbal products. When we started digging for information and researching product availability, one name kept cropping up: Dr. Khem Shahani, professor of Food Sciences and Technology at the University of Nebraska. It wasn't long before Siegfried and I were on our way to Nebraska to learn from the authority on bacteria, good and bad.

The Good Guys . . . .

The definition of good bacteria is simply that it contributes to the maintenance of the host in a state of good health. To survive within the caverns of our bodies, good bacteria must ensure that a healthy internal ecosystem is developed and nurtured. The good bacteria produce various vitamins, enzymes and natural antibiotics. They do this to create their own food and other materials that make their work easier to perform. As a pleasant side effect, the host derives huge benefits from the same vitamins and enzymes. To police their home turf, good bacteria keep the intestines clear of garbage and toxins and wage war with any

To police their home turf, good bacteria keep the intestines clear of garbage and toxins and wage war with any invaders.

Bacteria play a vital role in our well being, and strive to keep our internal Garden of Eden lush.

invaders, be they bad bacteria, viruses, poisons that may have entered with food, or yeasts that are out of control. If your body is not abused by junk food or environmental toxins, your friendly bacteria will work tirelessly on your behalf.

Friendly bacteria are the workers that strive to keep our internal Garden of Eden lush and productive. They are the farmers that cultivate and process raw materials, they are constantly at work to keep our garden weed free, they transport nutrients to access sites, they keep our streets free of garbage and they are the fearless soldiers that do battle with anything that threatens their host. Friendly micro-organisms follow nutrients from start to finish without once asking for time off or threatening to go on strike.

Proven Benefits of Friendly Bacteria

Worldwide research into the activities of various bacteria within the human body has proven the critical role that bacteria play in maintaining good health in the human organism. These benefits are summarized below.

- The number-one benefit of beneficial bacteria, according to controlled studies, is that they keep potentially patho-genic (bad!) bacteria from establishing a base in the colon.
- Good bacteria play an essential role in helping us digest proteins, good fats and carbohydrates.
- They have the ability to manufacture vitamins A, B and K and the natural antibiotics hydrogen peroxide and lactase.
- They help control excess levels of LDL cholesterol (the bad kind).
- They help prevent the formation of some cancer-causing substances in the body, and actively fight some types of tumors.
- Good bacteria actively support the immune system.

- They constantly battle and often destroy potentially dangerous and pathogenic bacteria and even viruses.
- They are effective in treating a variety of conditions relating to vaginitis and genito-urinary tract problems, and in controlling yeast infections, including athlete's foot, thrush and Candidiasis.
- They play a key part in assuring normal regularity and have proven effective against diarrhea and food poisoning.
- Friendly bacteria help maintain normal levels of fertility by balancing sex hormones.
- They even contribute to healthy skin while controlling various skin infections, including acne.

Which Ones Are the Good Bacteria? . . . .

There are only a few friendly bacteria that you should concentrate on. Forget about the bad bacterias for now; we will look at them in more detail later. The most common bacteria classed as beneficial to warm-blooded creatures like us are *Lactobacillus acidophilus*, *Bifidobacteria bifidum*, *Streptococcus thermophilus* and *Lactobacillus bulgaricus* (a transient but very good bacteria). To simplify, we will use the initial "L" for all lactobacilli, "B" for all bifidobacteria and "S" for all streptococci. While there are many supportive bacteria, the ones that are crucial in a supplement are *L. acidophilus*, *B. bifidum*, and *L. bulgaricus*. We'll look at the importance of each later.

Lactobacilli are found in milk products because they live in the grasses eaten by cattle.

The first and most easily recognized of the friendly bacteria are the lactobacilli. These bacteria are everywhere. The name is derived from lacto, meaning "found in milk" and bacillus, which denotes bacteria. Lactobacilli are found in milk products because they live in abundance in grasses that are eaten by cattle. Think about the different kinds of natural cheeses that come from around the world. Cows in different places eat different kinds of grass, and these various grasses play host to different bacteria. The natural cheeses are different because the bacteria are different.

The lactobacilli are but one family of bacteria. The family name covers many different types and strains (more than 200), but the one that is most beneficial and dominates within the human body is L. *acidophilus*. As the name suggests, it thrives in the slightly acidic environment within our intestines. L. *acidophilus* resembles links of sausages; there are about twelve links in a mature bacterium. They multiply, under ideal conditions (in a lab), every twenty minutes: one becomes two, two become four and so forth. In the human body, however, it is a constant battle for bacteria to survive and complete their many complex tasks. Under these circumstances their ability to multiply may be limited to once every twelve to twenty-four hours.

L. *acidophilus* lives and works in the bottom two thirds of the small intestines, the mouth, the colon and, to a lesser degree, the vagina. Bifidobacteria (there are four known strains) are found predominately in the colon and the vagina. L. *bulgaricus* is a transient bacteria that, although it does not set up housekeeping in our bodies, has a profound positive influence on our health. While there are many additional resident and transient good

B. bifidum is the dominant healthy bacteria in children under seven years of age.

12

bacteria working within us, they are really supporting players and are not often considered for supplementation.

In some instances one type of bacteria plays a more important role than the others. For example, the dominant bacteria of newborns, nursing mothers, children less than seven years of age and total vegetarians is B. *bifidum*. In everyone else the dominant good bacteria is L. *acidophilus*. The reason for this is based on the requirements of the individual at a specific time. When a child is born via the birth canal it is coated with mucus that is teeming in B. *bifidum* bacteria, the first line of defense against the outside world. Children born via Cesarean section are deprived of this protective bacteria and often develop early health problems that require antibiotic intervention, thus setting in motion the potential destruction of the child's fragile internal ecosystem.

The Bad Guys . . . .

The world within us is truly a wonder to behold; all is peace and harmony. Truly, our Garden of Eden is safe and secure...but wait, what is that lurking in the corner? In our tiny perfect world, could it be that there are also bacteria that might do us harm? Aren't all bacteria the same? I'm afraid not. Just as there are good people and bad people in the world, there are good and bad bacteria in the world within us. Only by being aware of the differences can we appreciate the role played by the trillions of good bacteria that work within our internal gardens.

The bad guys are not nearly as straightforward as our friendly bacteria. These invaders include more than forty common types of bacteria, along with countless other microorganisms such as viruses, yeasts and fungi. Just like our friendly bacteria, these unwelcome intruders are with us from the day we are born. Many of these bacteria and yeasts feel quite at home within our bodies and can't understand what we have against them. They would like nothing better than to have our good bacteria and immune system simply ignore them and let them live in peace.

To give an example of how hard the good guys have to work to keep the bad guys in line, it takes more than a million L. *acidophilus* bacteria to keep one yeast cell under control. If the

All is well when bad bacteria are kept under control by healthy levels of good bacteria.

numbers of L. *acidophilus* drop as a result of antibiotic or cortisone therapy, we are at risk of developing a yeast overgrowth. When bad bacteria are kept under control and harmless by healthy levels of good bacteria, all is well. But if the delicate balance is upset, the bad bacteria become pathogenic (worse than bad), producing toxins that could prove fatal to the host–this means you.

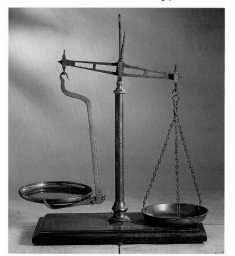

When the balance of good and bad bacteria is upset, bad bacteria become pathogenic–producing toxins.

Fortunately, the good guys have allies. Minerals play an important role in controlling some types of viruses and bad bacteria. For example, it was discovered several years ago that an over production of toxins by the bacteria *Staphilococcus aureus* in the vaginal tract caused toxic shock syndrome. Apparently, super-absorbent tampons, when used on light-flow days, continued to pull fluids from the area and as a result, levels of magnesium that usually controlled S. aureus were reduced to the point where the bad bacteria multiplied and produced high levels of very dangerous, and sometimes fatal, toxins. The result was toxic shock syndrome.

Our body is a very intricate and complex living machine. It depends on a carefully balanced but fragile ecosystem. We must keep in mind what is required of us to make this marvelous machine operate efficiently. Proper fuel and periodic maintenance is essential–you know, lube and oil and don't forget to change the filters. And of course, you must never put diesel fuel in your gas engine!

Antibiotics: Good, Bad or Ugly? . . . .

As children, many of us were exposed to one of the wonders of the twentieth century: lifesaving antibiotics. Penicillin, discovered by Alexander Fleming, was hailed as the answer to all health problems, a true panacea. But even Fleming realized that his discovery was a double-edged sword and could also cause irreparable damage as well as save lives. He cautioned his fellow scientists that penicillin, used indiscriminately, could severely damage the fine balance of our delicate internal ecosystem. He

14

also cautioned that some strains of strong, pathogenic bacteria would survive the penicillin therapy, adapt and become resistant to antibiotic treatment. How right he was; now we must suffer the consequences.

For every good bacterium we have working for us, there is at least one corresponding bad bacterium trying to make a home in our garden. Just like bad weeds, they seem to flourish with a minimum of help. These bad bacteria are responsible for everything from a sore throat to cholera. Often it is not the actual bacteria that cause the problem but the toxins they create. In their haste to control these bad bacteria, physicians are quick to prescribe a host of new, more potent antibiotics to seek out and destroy the bad guys. But we must realize that when this happens *the good guys are also fair game.* There is no place to hide. Antibiotics do not play favorites. They are unable to distinguish between good and bad: in the eyes of an antibiotic, a bacterium is a bacterium, and as such must be destroyed.

Invisible Antibiotics

Thank goodness I don't take antibiotics, you say. You might be surprised to learn that you have been taking small quantities of antibiotics for years, indirectly. Since the benefits of penicillin were discovered, scientists have looked for ways to capitalize on these benefits. As a result, widespread use of antibiotics was introduced into our food chain. Antibiotics were added to feed to reduce the risk of diseases spreading to people through beef, pork, fowl, eggs and processed foods. It was noted that livestock whose feed included antibiotics put on extra weight with less fat. Wow! Talk about a pleasant side effect. The belief was that if a little more was added, we could improve the yield even more. The rest is history, as more and more antibiotics found their way into our food, with predictable results: a decline in our friendly bacterial colonies.

15

You might be surprised to learn that you have been indirectly taking small quantities of antibiotics for years.

You might be wondering whether our friendly bacteria are also mutating into super bacteria, as the bad ones are doing, as a result of antibiotic use. With all these antibiotics, surely our internal garden should be in great shape, full of bigger and better good guys? The answer to this is a resounding "no." The good bacteria that survive the antibiotic battering now have to cope with viruses, which are immune to antibiotics; with yeasts that no longer are controlled by friendly bacteria and spread throughout the body, thumbing their noses at the guards; and with bigger and badder bad guys. Our friendly flora must struggle to survive and can't keep up. Toxins normally handled with ease build up and garbage and weeds are now found throughout our garden.

Antibiotics are in many instances lifesavers. Used carefully and only when absolutely necessary, they are very beneficial. Unfortunately, this careful and controlled use has not been the norm in the past. The prescribing of antibiotics has been subject to abuse. They have been used prophylactically ("just in case") for conditions such as chronic earaches and as a precaution any time minor surgery, such as the extraction of a tooth, has been performed. Antibiotics were routinely prescribed for viral infections simply because a patient insisted that they needed "something" for that rotten feeling. While they might have had a placebo effect on viral infections, that was the limit of their usefulness. Antibiotics do not harm viruses, period.

Whenever antibiotics are administered, for any reason, we must be aware that our friendly bacteria are under siege and need replenishing.

16

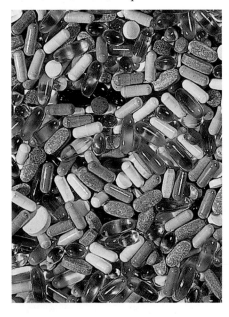

The Antibiotic Roller Coaster

The negative effect of antibiotics is not a one-time event, either. It can start a vicious circle: You take an antibiotic for a minor infection, it kills the infection (except for a few really strong bad bacteria) and it kills your friendly bacteria. The surviving bad bacteria eventually cause another infection, and because your defending colony of good bacteria is depleted, the infection is worse. So you take another, possibly stronger, antibiotic.

This kills more good bacteria and more bad bacteria, leaving a few even stronger bad bacteria floating around. Your defenses are down, and you get sick again. We all know people who are on the antibiotic roller coaster. Unfortunately, children are often subjected to this poor medical practice when they suffer from ailments such as ear infections.

The unpleasant side effect of this shotgun distribution of antibiotics is the irreparable effect antibiotics have on the friendly bacteria that are crucial to our very existence. Whenever antibiotics are administered, for any reason, we must be aware that our friendly bacteria are under siege and need our help. Many will not survive the course of antibiotics. It is therefore essential that we boost the friendly bacteria with foods and juices that will strengthen and feed them. The next step is to replace the ones that have been lost with a therapeutic-strength, high-quality bacterial supplement. The bacterial boosters are covered later on in this book, as are bacterial supplements.

Children are often victims of the antibiotic roller coaster, which results in recurrent ear infections.

17

Probiotics As Antibiotics

Good bacteria themselves have antibiotic properties. If our internal colonies of friendly flora are thriving, it is harder for the bad guys to get a foothold. All good bacteria produce acids, and some produce hydrogen peroxide; these greatly discourage bad bacteria. The healthier your internal colony, the less likely you are to ever need an antibiotic.

Treatment and Prevention with Friendly Bacteria

Controlling Yeast Infections

One of the major side effects of overuse of antibiotics is an over-growth of yeast, especially *Candida albicans*. Much has been written about so-called yeast infections. Many medical professionals do not even accept it as anything more than a "women's nuisance"

One side effect to overuse of antobiotics is yeast infections.

that comes around about once a month and goes away after their menstrual cycle is finished. This is, however, just the tip of the iceberg.

Yeast/fungi overgrowth or infection can be a daily nightmare that controls the lives of its victims. This insidious problem is shared by men and children as well. In addition to the best-known yeast infection, vaginitis, C. albicans also can cause inflammation of the mouth and tongue (called thrush, this is common in infants), inflammation of the rectum, and mental symptoms such as anxiety, irritability and depression.

Yeast is a form of plant life, a subgroup of the mold/fungi group. These microscopic plants are rootless and absorb their nutrients from organic material, dead or alive, that surrounds them. They offer no benefit to the host. The most common yeast in humans is Candida albicans: just about everyone has been colonized by this yeast—it is found on our skin and in the lower bowel and vagina. Normally, when it is kept under control by friendly bacteria, C. albicans is harmless. However, when it is allowed the opportunity to get out of control, such as when you have been using antibiotics, cortisone or hormone medication (such as the birth control pill) for a prolonged period, it transforms to its fungal form, puts down roots (rhizomes) and can then spread quickly throughout your body.

These rhizomes can puncture the gastrointestinal lining and allow toxins into the bloodstream. Your immune system is then called upon to fight these toxins. When this happens again and again, your immune system can't help but be drained. The transformation of C. albicans from yeast to fungus is prevented by biotin, which is produced by your colony of friendly bacteria.

You have only to look at the number of advertisements touting quick-fix "cures" to realize that yeast/fungal infections are all too common. Contrary to being strictly a localized nuisance problem, yeast and fungal infections that have been treated with ineffective antifungal "cures" can eventually spread to every organ in the body, including the brain. Thank goodness there are natural champions in this struggle to control an intruder that has been with us since birth.

Those champions (along with a daily routine of candidiasis-fighting supplements) are the friendly micro-flora that inhabit the breeding ground and home base of yeast and fungi, namely, the colon: L. *acidophilus* and B. *bifidum*. The task is not an easy one, especially if the yeast has mutated into the mycilial (rooted) fungal form that can and does attach itself like a parasite to every organ within us. Once it has penetrated the brain's protective barrier, the saying "it's all in your head" takes on a far more ominous meaning. It will take supplements with special talents to chase the fungus at this stage: in the colon, it takes one million of the super strain of L. *acidophilus* to control just one yeast cell.

The diet to control overgrowth excludes sugars of all kinds, including fruit, alcohol, and caffeine.

To control yeast infections, supplement therapy should be used in conjunction with a strict candidiasis diet. This diet excludes sugar of all kinds, including natural fruit sugar; alcohol; refined flour; food coloring, flavoring and preservatives; caffeine; processed meat; and all fermented and yeast-derived foods. For complete and easy-to-understand information on yeast infections read *Nature's Own Candida Cure* by William G. Crook MD (*alive* Natural Health Guides #18, 2000).

While your friendly bacteria are working to get the yeast overgrowth under control, it is normal to feel less than your usual self for about a week. This is a result of your body processing the dead yeast. Don't stop treatment because of this under-the-weather feeling. In fact, if you have had an ongoing candidiasis problem, you should continue this intensive treatment for at least six months.

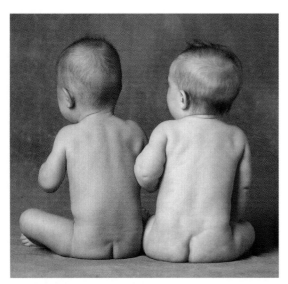

Special attention should be given to the role friendly bacteria can play in infants and the growing incidence of yeast infections.

Controlling Yeast Problems in Infants: With the growing numbers of newborns and infants developing yeast infections shortly after birth, I feel special attention should be given to the role friendly bacteria play at a very early stage in life and the long-range benefits that can be obtained from using bacterial supplements as soon as they are needed.

Let me give you an example. When my granddaughter was born, her mother had a very difficult labor; after some twenty hours, her doctor decided that a normal delivery was becoming dangerous and performed a Cesarean section. Due to complications that are normal with surgical intervention, both mother and daughter were administered antibiotics. As a result, all the friendly bacteria (already minimal in the baby as a result of the C-section delivery) were compromised. The infant developed a candidiasis diaper rash.

The pediatrician, a very dedicated woman, prescribed a nystatin cream, but the rash persisted. At this time it was also discovered that the mother was not able to breastfeed her child, thus depriving the baby of essential B. *bifidum* bacteria found in mothers' milk that would have fought the C. *albicans* rash.

I suggested mixing the super-strength bacterial supplement (the DDS-1 strain, which I'll tell you about later) of L. *acidophilus* with plain yogurt and applying that to the rash. I also told the mother about the other benefits of adding the supplement to the baby's formula to give the child's internal colony a boost, but her parents assumed that the pediatrician knew more than Gramps, and did not heed my advice. I mentioned that it was common practice to prescribe a stronger cream if the one being used was not effective. My prime concern was the possibility of a cortisone cream being used in a delicate area. Conceivably, if used too liberally, this can cause sterility in the child.

As I predicted, a new prescription was given. Fortunately, the pharmacist cautioned against overuse of the cortisone cream, indicating that it could cause damage if not used very sparingly on an infant. At this point, the baby's mother decided to try the yogurt/*acidophilus* mix instead and within three days the candidiasis rash was completely gone. She took my other suggestion to heart, and added the DDS-1 strain of acidophilus to the baby's formula. The child has remained yeast free.

Breastfeeding, Formula and Friendly Bacteria: To determine the benefits and drawbacks of each feeding method, we must look into the different factors that influence nutrition and friendly bacterial levels. The main benefits a child receives from mother's milk are the natural human fats and balanced nutrients that are essential to healthy growth. Mother's milk is the only source of GLA-gamma linoleic fatty acids for infants. It also contains many antibodies to ward off bad bacteria and other unwanted intruders that the infant might come in contact with in the potentially hostile environment in which we live. There is another plus that is of equal importance to the antibodies within the milk, and that is the friendly *B. bifidum* bacteria that are transferred from mother to child.

We must start the infant off with the best chance to remain disease free. The best way to do this is with the natural antibodies and friendly bacteria found only in a healthy mother's milk. We have all been exposed, to a certain degree, to a collage of chemicals and antibiotics in all facets of the environment. The antibodies that a woman's body has built up to this increasingly hostile environment could prove crucial to the long-term health of her infant. Mothers' milk is also high in lactose and low in protein (particularly casein). As well as supplying living *B. bifidum* bacteria, it further stimulates the growth of friendly bacteria already in the child's system, thus maintaining a healthy, slightly acidic pH level in the child. Human milk contains essential forms of acetic, lactic and formic acids, as well as large

Infants have the best chance to remain disease-free when they begin life with the natural antibodies and friendly bacteria found only in a healthy mother's milk.

21

numbers of macrophages, which are cells that seek out and destroy potentially harmful bad bacteria.

While breastfeeding is more beneficial than any formula, we must also be aware of research that is going on worldwide into environmental influences on each of us, and in particular how these factors have an impact on a nursing mother's ability to supply well-balanced nutritional support to her infant. Research over more than twenty years has shown that the number of bifidobacteria in breastfed infants has declined, while the presence of harmful *Escherichia coli* bacteria has increased steadily, along with other disease-causing bacteria. Of prime concern is the revelation that some of these strains of E. *coli* have become resistant to such antibiotics as neomycin, one of the more popular and effective antibiotics.

When breastfeeding is not possible, formula is an acceptable substitute; however, there will always be some very beneficial components missing. Since formulas are based on cow's milk, the structure is different, since the requirements of humans and cattle do, of course, vary substantially. The difference is obvious in the infant's stools. In breastfed babies, the pH level is slightly acidic (around 5) and the stools are very soft with a cheesy odor. The stools of formula-fed babies have a less-acidic pH level of between 6 and 7 and are heavier and foul smelling. For infants who must rely on a formula, I strongly recommend that parents consider adding a quality bacterial supplement on a daily basis. It is important to ensure the supplement contains the B. *bifidum* bacteria as well as the DDS-1 *acidophilus* strain.

Fighting Cancer

Tiny though they are, your resident friendly bacteria can help you prevent and fight one of the most frightening illnesses there is: cancer. Let's look at prevention first.

Before we even talk about how bacteria supplements can help you, you must understand that maintaining a healthful diet is crucial to preventing many cancers from ever getting a foothold in you body. It has been estimated that 35 percent of cancers are diet-related, and animal fats and proteins are the biggest culprits. These foods do two things: they cause putrefaction in the gastrointestinal tract, which leads to a decrease in friendly bacteria and an increase in the bad ones. They also encourage this increased bad population to produce certain enzymes, and these enzymes can convert the chemical products of digestion into carcinogenic chemicals.

In addition to severely limiting these elements in your diet, you should increase the amount of cultured milk products and fermented foods that you consume. These life foods increase the level of lactic acid in the intestines, and this helps get rid of dangerous micro-organisms, which prefer an alkaline environment. Populations that eat a lot of this type of food have low rates of cancer.

"If the Germans ate as much sauerkraut as the French believed they did, there would be no cancer in Germany!"
—Johannes Kuhl, MD, PhD

How can supplements help? Research on humans has shown that L. *acidophilus* has the ability to inhibit the formation of carcinogens by switching off the enzymes produced by harmful bacteria that cause the formation of carcinogens. Let's look at an example. You are familiar with nitrites, which are found in abundance in processed meats and are something to be avoided in the first place. If you do eat them, however, bad bacterial enzymes in your body can convert nitrites into nitrosamine, which is a cancer-causing (carcinogenic) substance. Good bacteria neutralize these enzymes, so that nitrites stay nitrites. In addition, good bacteria have been shown to actually eliminate the nitrites in the first place, so that the bad bacterial enzymes don't even get a chance to change it into nitrosamine. The amazing super stain of L. *acidophilus* even has the ability to convert cancer-causing substances such as nitrosamine back into their less-dangerous forms.

Don't think that by taking a bacterial supplement you can eat tons of red meat and fatty foods with impunity! An animal study has shown that the bad effects of eating too much animal fat persist to some extent, even when L. *acidophilus* is supplemented. Good as it is, it is not a silver bullet; you need to adjust your diet, too.

Taking on Tumors: If you have cancer, friendly bacteria supplements could help. Not only do these amazing little creatures help to prevent cancer in the first place, studies have shown that they can actually help to dissolve some tumors. L. *bulgaricus*, so named because it was discovered in 1956 in Bulgaria by Dr. Ivan Bogdanov, synthesizes an anti-cancer substance that kills tumor cells, without harming the patient. Dr. Bogdanov discovered that the active component in L. *bulgaricus* was peptoglycan, which is found in the cell walls of the bacteria. In fact, this compound is found in all the lactobacilli. Dr. Shahani found the greatest cancer-fighting effect in a study on mice using the DDS-1 super strain of L. *acidophilus*.

Not only does L. *bulgaricus* work directly on tumors by dissolving them, but because it is not a resident bacteria, it also stimulates the immune system, which is important in fighting cancer. In addition, Dr. Bogdanov found that treatment with L. *bulgaricus* had a therapeutic effect for patients suffering from the negative effects of radiation and chemotherapy, and in fact helped to prevent these negative effects in patients about to receive these powerful and potentially debilitating therapies.

Cancer treatment with friendly bacteria is intensive and long term, but Dr. Bogdanov found that in some cases it was effective even in patients who had been declared terminally ill. The treatment, which lasts at least six months, consists of a highly concentrated substance, called Anabol, extracted from L. *bulgaricus* synthesized products and processed cell bodies. A 10-15 gram daily dose of Anabol contains as many L. *bulgaricus* cell bodies and as much synthesized anti-tumor product as 40-60 kilograms of Bulgarian yogurt! Huge doses of 30-40 grams have been shown to actually dissolve tumors, but this aggressive treatment is hard on critically ill patients, because the dissolved tumor material is released into the body, and the already overwhelmed immune system has trouble coping with it. Gradual treatment is preferred.

Dr. Bogdanov's studies in the 1960s showed that in patients who were terminally ill with a variety of cancers, some experienced complete regression of their tumors, and most had some improvement. Importantly, none had negative side effects. In

Not only do these amazing little creatures help to prevent cancer in the first place, studies have shown that they can actually help to dissolve some tumors.

1989, a summary of six studies confirmed that lactobacilli can suppress some tumor activity.

Colon Health for Overall Health

The colon is the final section of the intestines. It starts at your lower right side, travels up to just under the rib cage (ascending colon), cuts across to your left side (transverse colon), takes a sharp turn downward (descending colon), and gradually forms a storage pocket (sigmoid colon) before forming an exit (anus). The colon extends approximately three to four feet in length and comes in contact with most of our internal organs at one time or another. Is it any wonder that if a person develops cancer of the liver, pancreas or stomach, its source often can be traced back to the colon?

Unless we maintain good colon health through efficient elimination of toxins, ensured by daily regularity, then some form of disease is not only possible, but inevitable. The only way this can be accomplished is if we maintain a healthy ratio of friendly bacteria to bad coliform bacteria. It is the responsibility of the good bacteria

When the bowel is not healthy, the rest of the body suffers a similar fate.

First Light

to ensure the manufacture of vitamins, enzymes and a host of other essential elements within the colon.

When the balance of good versus bad is maintained, all else will operate normally and regularity will be assured. This need for regularity is obvious when you consider that we reabsorb up to two liters of nutrients from the colon every day. It is imperative that we keep toxic materials moving out of our bodies, thus reducing the likelihood of reabsorbing these life-threatening toxins along with the nutrients. Since prevention is always preferable to damage control, we must do whatever it takes to maintain a healthy colon.

What Goes On in the Colon? The food we eat travels through some twenty-one feet of small intestines and is exposed to various enzymatic actions that extract nutrients at certain sites along the way. Your body is a marvelous piece of machinery and the intestines are like a fuel line. This fuel line is designed with tiny hair-like tunnels running off it that allow us to extract materials as the nutrient mass moves along. As this food travels lower in the intestines, the tunnels leading off become larger, allowing larger nutrients to be utilized. When the residue of unused matter reaches the end of the small intestine, a valve opens and allows it into the large intestine (colon).

This residue has had a great many essential nutrients removed, but not all. There is still a lot of activity involved before your system is finished with this material and ready to move it out of the body as waste. This is the area where our friendly bacteria are most active, where they really get to work. Enzymes get down to business, bacteria deal with invaders and toxins, yeasts are told to behave, vitamins are produced and all is well, unless of course your colon is under stress or out of balance. For example, an overgrowth of E. coli could create toxins that could severely upset the delicate balance within the colon.

According to controlled studies, certain strains of E. coli produce a substance that the body cannot distinguish from insulin. This substance appears to form a block at special receptor sites in cells that real insulin needs to reach to do its job of controlling blood sugar levels. In a healthy colon, this bad bacteria would be controlled by the friendly micro-organism B. bifidum.

How Does Bacterial Activity in the Colon Determine Your Health? Friendly and unfriendly bacteria are in a constant battle for dominance throughout your body. Each tries to establish colonies along the inner walls of your intestines. In the more acidic areas, the odds favor the good guys. But if they let their guard down, or if their numbers are reduced by a poor diet, stress, or broad-spectrum antibiotic therapy, then the bad guys (pathogenic bacteria, viruses and yeasts) are quick to jump in and start creating levels of toxins that wreak havoc with your immune system and your total state of health. These snakes start to take over in your internal Garden of Eden.

The area that is most susceptible to these invaders is the large intestine (the colon). This is the critical area where waste material is deposited before it is slowly moved out of the body. A healthy colon contains 85 percent friendly flora and 15 percent coliform and putrefactive bad bacteria. When this balance is upset, fermentation and putrefaction set in, regularity suffers and there are dangerous delays in clearing toxic materials out of your system. Without the trillions of friendly micro-organisms, your colon stops extracting essential nutrients and becomes a cesspool of putrefaction. When the bowel is not healthy, the rest of the body suffers a similar fate. We can develop skin eruptions, allergies and migraines, with the possibility of more serious diseases to follow. Conversely, when the colon is in good health, with high levels of friendly bacteria, then the rest of the body will also be in good health. A healthy colon means a healthy body.

Corina Messerschmidt

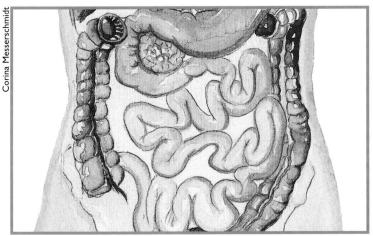

By being aware of the danger signals that are as obvious as chronic diarrhea or constipation, you can take corrective measures. The single most important remedial or preventive measure you can take is to maintain maximum levels of friendly micro-flora within your colon, either by boosting existing good bacteria or supplementing with quality, therapeutic-strength bacterial supplements.

The Immune System

The immune system is highly dependent on our maintaining a healthy level of good, friendly bacteria. This is a prime example of a true symbiotic relationship. When bad bacteria, viruses and yeasts get the upper hand, the immune system is placed under extreme stress. In addition to its regular work protecting you, your immune system now must devote itself to battling factors to which it is not accustomed, intruders that are normally taken care of by your community of friendly micro-flora, L. *acidophilus* and B. *bifidum*. But these good guys are themselves fighting just to survive the antibiotics, the fast-food fats, the excess of sugars, tars, trans fatty acids and a host of other hard-to-assimilate materials.

Antibiotics are useless against viruses, such as the flu virus, and kill off friendly bacteria.

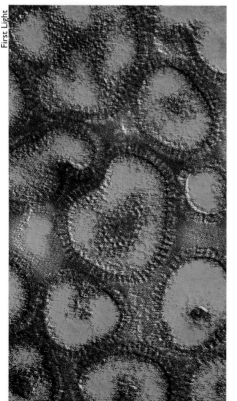

When the levels of good bacteria drop to a critical point, our immune system finds itself unable to cope and sends out conflicting signals. We start to have symptoms of everything, and nothing. We know there is something wrong with us but can't seem to pin it down. If we are smart and lucky, we may consult a natural health professional who will use the symptoms to lead him or her to the true source of the problem and to determine what remedial action is indicated. A less happy scenario would be a visit to a medical doctor's office. With so many and varied symptoms, a series of tests will be performed, antibiotics might be prescribed and the merry-go-round will continue.

Once the balance between the good and bad tips to the bad, prompt remedial

28

action is needed. Bad, out-of-control bacteria are known to be responsible for everything from a sore throat to the bubonic plague. The bad bacteria is aided and abetted by myriad viruses, many causing symptoms similar to those caused by bad bacteria. When this happens, antibiotics might be prescribed. Viruses really love it when this happens, since antibiotics are useless against viruses. The virus can therefore thrive while the good bacteria that should be controlling it are being attacked and destroyed by the antibiotic. The end result can be catastrophic.

When bad bacteria, viruses, and yeasts get the upper hand, the immune system is placed under extreme stress.

Helping Your Liver

In today's atmosphere of water-, air- and food-borne toxins, friendly flora are more essential than ever to maintaining a vibrant and healthy liver. After your skin, your liver is the largest organ of your body. While the liver is very efficient at filtering out toxins from our blood, it has a very busy schedule and can use all the help we can give it.

Functions of the Liver

To appreciate the role that friendly bacteria play in supporting the crucial cleansing functions of our liver, it is important to note just what the liver's functions are. The liver filters impurities and toxins out of our bloodstream and it converts amino acids into various other components that contribute to the production of essential protein elements. The liver produces anticlotting and anticoagulating factors, it is the source of red blood cells in the growing fetus of expectant women and it is the main site for the production of plasma proteins.

The liver is the warehouse of the body: most of the storage space is taken up with good stuff such as vitamins B1, A, D, E and K. It also plays a key role in controlling body heat by regulating blood volume. HDL cholesterol (the good one) is also produced in the liver. Whew! Talk about a workhorse! Is it any wonder that periodically our liver needs a little help? For complete and easy-to-understand information on this subject read *Liver Cleansing Handbook* by Rhody Lake (*alive* Natural Health Guides #4, 2000)

Help Is on the Way: When the liver is overexposed to toxins produced by E. coli and other pathogenic bacteria, it could develop very serious problems if it doesn't get help. One of the problems with which we are familiar is cirrhosis of the liver. To ensure that these toxic time bombs are defused, it is essential that we work hard at maintaining healthy levels of internal guardians, our friendly L. *acidophilus* and B. *bifidum* bacteria. The most beneficial friendly bacteria for our liver is B. *bifidum*. It plays a key role in the colon neutralizing and reducing the levels of ammonia in the bloodstream. Excessive levels of ammonia that enter the liver via the bloodstream can cause a toxic overload that can eventually adversely affect normal brain function. Once again, a healthy colon means a healthy body.

The effect of friendly flora on the liver has been the subject of medical research. In one study, patients with cirrhosis of the liver had an average level of lactobacilli of one million bacteria per gram of bowel contents, and 100 billion per gram of the bad bacteria E. *coli*. Not a good ratio! These patients received supplements of lactulose, a substance formed from the milk sugar lactose. Lactulose is extremely beneficial to our colonies of B. *bifidum* bacteria, which in turn help our liver in its detoxifying work. After supplementation with lactulose, the level of lactobacilli in these patients rose to between 10 and 100 million per gram of bowel contents, and the level of E. *coli* had fallen to almost this same point, a dramatic improvement of the good guy-bad guy ratio.

While lactulose itself is not available as a supplement, the results of this study underline the importance of keeping your colony of B. *bifidum* strong and healthy, and this can be aided by supplementation with B. *bifidum* bacteria themselves.

Help with Your Cholesterol Level

Research has shown that L. *acidophilus* can lower cholesterol.

If you're like many people in the North America, you might be trying to lower you blood cholesterol level. There are two kinds of cholesterol: high-density lipoproteins (beneficial; it helps to reduce the bad kind) and low-density lipoproteins (dangerous). The LDL cholesterol builds up in the arteries and contributes to heart disease. Some cholesterol is consumed in the diet, but most is manufactured by your body. The more animal fat you eat, the more cholesterol your body makes to attempt to break it down. Cholesterol is one of the raw materials used by the liver to make bile, which is needed to break down fat. Your body is environmentally friendly: it recycles bile, but to do this it needs a healthy colony of B. *bifidum* working with it in the intestines.

You are not alone in your battle against cholesterol: you have millions of friendly flora to help you. Research has shown that L. *acidophilus* can lower cholesterol levels. But not just any *acidophilus* will do. For example, one study used two strains: RP32 and P47. RP32 was able to break down cholesterol in the intestines, but P47 was not, so you must choose carefully. The DDS-1 super strain has been shown to produce the anti-cholesterol effect.

The L. *acidophilus* in Kefir (a fermented milk product) produces an anti-cholesterol effect.

L. *acidophilus* in milk (the effect has also been demonstrated by live yogurt) produces something called the anti-cholesterol factor. This is found in fermented milk products like natural yogurt, quark, butter, milk, and kefir, but also in sweet (unfermented) acidophilus milk. It is this anti-cholesterol factor that has the ability to reduce the level of bad (LDL) cholesterol in your blood. In addition, one study showed that acidophilus actually seemed to absorb cholesterol when the acidophilus was grown in a cholesterol-rich environment. So friendly bacteria help with cholesterol control in two ways: B. *bifidum* works in your intestines to aid in the breaking down and recycling of bile, which is what breaks down fats, and L. *acidophilus* has the ability to directly absorb cholesterol.

Some cultures, such as the Masai of Africa and the Mongols of northeastern Asia, consume large quantities of saturated animal fat, and still have low cholesterol levels. Why? They also consume vast quantities of fermented milk products. Similarly, the amounts of milk or yogurt used in studies are sometimes huge. You might not be able to change your diet that drastically; a good-quality supplement containing the DDS-1 strain of L. *acidophilus* is a better bet.

Skin Conditions

Acne and herpes simplex infection are just two more conditions that can be treated with lactobacilli.

When combined with fresh aloe vera, L. bulgaricus powder makes an effective acne ointment.

Acne: A doctor in Baltimore, Maryland, who treated his patients' gastrointestinal problems with L. *acidophilus* and L. *bulgaricus* tablets, found that if the patient also had acne, this tended to improve. He looked at the results more closely, and found that of 300 cases, 80 percent experienced an improvement in their acne. The results were best in teenaged patients. Although the acne tended to get worse very briefly after treatment was started, in most cases improvement was seen within two weeks.

In Europe, a paste made from L. *acidophilus* has been used to treat acne for years. It is applied directly to inflamed areas. A similar product has been made with L. *bulgaricus* cultured in milk.

Make Your Own Acne Ointment

Combine 1 teaspoon of L. *bulgaricus* powder with enough fresh aloe vera (scraped from the inside of a leaf) and distilled water to form a smooth paste. Wash your face, pat it dry and apply the ointment to your face and neck. Leave it on for twenty minutes or more, then rinse off with warm water. Pleasant side effects: this treatment also acts as a gentle exfoliant and smoothes and softens skin. You might also find that it tightens pores and diminishes wrinkles!

Herpes Simplex: In the 1950s, a very forward-thinking doctor in Massachusetts was treating diarrhea with L. *acidophilus* and L. *bulgaricus*. In patients who also had herpes simplex sores in the mouth (cold sores), he noticed an improvement in these, too. He decided to investigate further, and treated more patients with mouth ulcers and patients with genital herpes. Of seventy patients with genital herpes, forty-three were cured of their outbreak and twenty-four were "much improved;" only three were unaffected by the treatment. The success rate for mouth ulcers was 80 percent, with improvements seen as soon as twenty-four hours after the start of treatment. This doctor did not claim that the herpes infection itself was cured, just that individual outbreaks could be quickly controlled. These results were repeated in other studies: in one, thirty-eight of forty patients enjoyed fast relief of mouth ulcers after treatment with lactobacilli.

The success rate for mouth ulcers was 80 percent, with improvements seen as soon as twenty-four hours after the start of treatment.

Bacterial Supplement Dosages

Condition	Bacteria	Dosage
General use	L. acidophilus	½ tsp daily
	B. bifidum	⅛ tsp daily
Acne	L. acidophilus, B. bifidum, L. bulgaricus	½ tsp each twice a day
Candidiasis	L. acidophilus, B. bifidum	½ tsp each at meals
	L. bulgaricus	¼ tsp after meals for a week or more
Cholesterol control	L. acidophilus, B. bifidum	½ tsp each at meals
Cold sores	L. acidophilus, B. bifidum, L. bulgaricus	1 tsp each before meals
Constipation	L. acidophilus, B. bifidum, L. bulgaricus	½ tsp each before meals
Diarrhea	L. acidophilus, B. bifidum, L. bulgaricus	½ tsp each every half hour
Formula-fed babies	B. infantis	Start at ⅓ tsp daily; increase gradually
Lactose intolerance	Milk-cultured L. acidophilus DDS-1	Start at ⅛ tsp daily; increase gradually
Liver detoxification	L. acidophilus, L. bulgaricus	½ tsp 3 times per day
	B. bifidum	2 tsp 3 times per day
Pregnancy/nursing	B. infantis	½-1 tsp daily before meals
Toxic shock syndrome	L. acidophilus, B. bifidum, L. bulgaricus	½ tsp each twice per day
Urinary tract infection	L. acidophilus, B. bifidum, L. bulgaricus	1 tsp each twice per day for 2 weeks
Vaginitis	L. acidophilus, B. bifidum, L. bulgaricus	1 tsp each twice per day for 2 weeks

All amounts listed are for powdered bacteria. It should be dissolved in 8 ounces of tepid, non-chlorinated water and, unless otherwise specified, taken forty-five minutes before meals. These amounts are for adults; reduce accordingly for children,* and always check with your health-care professional before using any supplement.

*A child of 10 or older may take half of the adult dose depending on weight for his or her age. It is always best to consult your natural health care provider for specific dosages for children.

Take Home Some Friendly Bacteria Today . . .

I can only imagine the confusion of the average person, with little knowledge of friendly bacteria, going into a store seeking help in choosing a bacterial supplement. The array before us is massive. The cooler is bulging with different products. There are bacteria in glass bottles and others in plastic bottles, there are powders, capsules, tablets and wafers; some come in a tasty liquid. Some are refrigerated, while others claim no refrigeration is required. There are milk-based and non-dairy-based products. The choice seems endless. To add to the confusion, most staff people, and even many health professionals, are in the dark and can offer little knowledgeable help.

The Crucial Component: L. Acidophilus

Some manufacturers just call their product DDS, while DDS-1 stands for the research document #1 at the Department of Dairy Science.

The most important good bacteria over the long haul is L. *acidophilus* with its 200 known strains working synergistically to maintain a state of good health in its host. Fortunately, pioneers in the field of bacteria research managed to culture and perfect single-strain friendly bacteria, such as L. *acidophilus* and B. *bifidum* that will perform most of the tasks that previously took hundreds of non-specific bacteria to perform. The new strain of L. *acidophilus* developed by Dr. Shahani and associates at the University of Nebraska is called DDS-1 acidophilus. It is the most effective against the greatest number of hostile micro-organisms. It is also a great help with nutrient absorption. Be sure this is the strain you are buying.

Unfortunately, L. *acidophilus* is the most-abused supplement offered by most suppliers. Non-specific acidophili are very inexpensive to culture, but they are the weak links—they need each other to perform even the simplest tasks and when the body is under attack they are the first to suffer. On the other hand, L. *acidophilus* is the bacteria most crucial to our well being. With this in mind, I always insist on the super DDS-1 strain of this bacteria.

The DDS-1 strain can perform all of the tasks that are normally performed by the 200 non-specific strains.

Over the years I have had occasion to work with many different bacterial supplements, but in the end I have eliminated the non-specific strains and concentrated on the bacterial supplements that have proven to have therapeutic qualities. There are currently two bacterial supplements available that contain this super DDS-1 strain of *acidophilus*. If the label does not list the DDS-1 strain, it contains one or more of the non-specific acidophili that, while they will do no harm, will have limited results for specific health problems.

Second in Command: B. Bifidum

The B. *bifidum* bacteria that were isolated and improved upon were not given specific, simplified names. In choosing a B. *bifidum* supplement I depend on the same supplier that provides my DDS-1 *acidophilus*.

For children up to age seven or eight who have been exposed to antibiotic therapies for chronic ear infections or other ailments, it is imperative that their already devastated ecosystem of friendly micro-flora be boosted with the best L. *acidophilus* with B. *bifidum* available. Forget about the other bit players that some suppliers may tempt you with; the two critical bacteria for children are L. *acidophilus* and B. *bifidum*.

The two critical bacteria for children are *L. acidophilus* and *B. bifidum*.

Just Passing Through: L. Bulgaricus

The other friendly bacteria that has proven to be extremely beneficial when you need a supplement for a serious problem is L. *bulgaricus*. This is a transient bacteria (just passing through your system). It will not set up housekeeping and is identified by your immune system as an intruder. In spite of this, L. *bulgaricus* helps produce lactic acids and lactase enzymes and seems to have a cleansing effect on the lymphatic system. As an intruder, it nudges a complacent immune system into action so that the immune system then attacks all intruders, including wandering *Candida albicans* cells that may be getting out of control.

A word of caution: L. *bulgaricus* is a strong, aggressive bacteria, much stronger than L. *acidophilus* or B. *bifidum*. Inside your body, L. *bulgaricus* will contribute to your well being without unduly affecting other bacteria; however, if you put it in a jar with other bacteria it will ultimately destroy the weaker bacteria it is in contact with. The only bacteria that tolerate one another in a jar are L. *acidophilus* (DDS-1 strain) and B. *bifidum*.

What About the Other Good Bacteria?

This is where the confusion really sets in. You will see a number of other bacteria listed on the labels of some supplements, each professing to be "essential" or "superior." Yes, all of these bacteria are good guys and yes, they inhabit our intestines and work very hard on our behalf, but they are not in the same league as L. *acidophilus* or B. *bifidum*. They are available for the sake of variety, to try to get the consumer to purchase more product. Additional bacteria you might find include L. *casie* (used in cheese making), L. *rhamnosis*, a sub-species of L. *casie*, L. *plantarum*, *Streptococcus faecium*, and so on and so forth ad nauseum. They range from having marginal benefits to being totally useless for therapeutic applications. The best description concerning these is found in the book *Probiotics* by Natasha Trenev and Leon Chaitow (Avery, 1998). Since I cannot put it any better, I will quote from them.

> Over the last few years, as more people have become aware of the usefulness of supplementation with friendly bacteria, especially certain strains of acidophilus, a marketing phenomenon has become evident. This can be termed the 'something-dophilus' phenomenon. There are now so many 'dophiluses' in health food stores and pharmacies, that the public may justifiably become confused as to which are sound and which are not. The price variations between these various 'somethingdophiluses' is enough to depress the potential purchaser, ranging as they do from fairly cheap to what appears to be exorbitant. This confusion is compounded by the label notes which often tell the buyer that the contents possess tens or hundreds of millions or even billions of active organisms per gram. How many do we need to do any good anyway? And more recently, to add yet further to the confusion, there have arrived a range of bacterial cocktails, usually containing a certain 'dophilus' content as well as a variety of other organisms (such as Streptococcus faecium and L. casie) almost always in unspecified quantities, except that the word "billion" usually appears somewhere on the label.

Is Ten Billion Not Better Than One Billion?

Much has been said concerning the apparently huge differences in the number of bacteria per gram in various supplements. How can we justify recommending a supplement that contains "only" one or two billion bacteria per gram when other less-expensive products claim such huge numbers, ten billion and more, per gram?

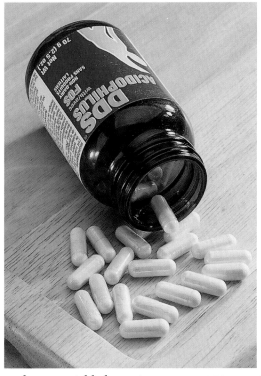

You may recall the structure of bacteria and the harvesting of these tiny micro-organisms. A mature lactobacillus is made up of approximately twelve links joined end to end. In the lab this is counted as one bacterium. If bacteria that have been cultured on solids are separated from their growing medium by means of a high-speed centrifuge, the links separate. Each link is then counted as one viable bacterium. One mature bacterium has become twelve fragile bacteria that will not reach maturity until they have grown to twelve links. But since they are now freeze dried and therefore dormant, it is virtually impossible for them to reach maturity until long after you purchase them.

The DDS-1 strain of acidophilus is harvested using a system that ensures the links of mature bacteria remain intact.

The DDS-1 strain of *acidophilus*, on the other hand, is harvested using a filtration system. This ensures that the links of the mature bacteria remain intact and each mature twelve-link bacterium is counted as one single, mature, viable bacterium that is ready for action as soon as you take it. To me it makes good sense to choose full-grown, mature bacteria over billions of fragmented bits and pieces. If you were building a brick wall, I am sure you would rather use whole bricks rather than try to piece together fragments of brick.

Powder, Capsules, Tablets, Wafers: Which Is Better?

In my opinion there are really only two choices: powder or capsules. The powder is usually more economical; however, some people feel that each time the bottle is opened and a spoon is inserted into the powder, contamination is possible. Others feel that the bacteria suffer somewhat during encapsulation and that this might adversely affect their usefulness. I believe that either form is acceptable. Powder is easier to mix with non-chlorinated water or yogurt for making colon or vaginal inserts, while capsules are more convenient, especially for traveling.

As for the tablets and wafers, I believe very strongly that the heat and pressure used in forming these will destroy live bacteria. They will be nutritious, but useless for supplementation. Dead bacteria do not contribute, period.

When deciding what form to buy your supplement in, keep the following in mind:

- Liquid bacterial supplements, usually found on the shelf and marked "no refrigeration needed," will make a nutritious drink, but will offer few living bacteria. The liquid will, however, provide nutritional support to the good bacteria already living inside your body.
- Flavored wafers are tasty but useless. Living bacteria simply do not survive the heat and pressure involved in forming the wafers.
- Tablets fall into the same category as the wafers.
- Supplements should be in dark colored glass bottles to filter out light and prevent moisture from affecting the slumbering bacteria. Some manufacturers use plastic to cut costs, but plastic is slightly porous and over time will allow moisture into the bottle, thus compromising the viability of the contents.
- To maximize the survival of the bacteria in any supplement once you consume it, it is recommended that the supplement be taken with tepid, filtered, non-chlorinated water and on an empty stomach. Water, with a neutral pH level of 7, does not trigger the production of stomach acids, which allows the water–along with the bacterial supplement–to pass through this potentially dangerous area intact.

Dairy or Non-Dairy Based?

Some manufacturers make a big issue of dairy-based versus non-dairy-based supplements. Both are good if the bacteria within are good. If, however, you start with an inferior group of non-specific "somethingdophiluses" with an assortment of other bacteria added for variety, then you will still wind up with a product that simply will not do the job you want it to do. If you buy a lesser-quality product and you are lactose intolerant, then it is important to choose a non-dairy-based bacterial mix; it will undoubtedly be unable to produce lactase. The advantage to insisting on a quality supplement that contains the super strain of *acidophilus* (DDS-1) is that you get bacteria that produce the enzyme lactase, which is essential to converting lactose within the digestive system. For lactose intolerant people, I recommend choosing the DDS-1 strain of *acidophilus* in a non-dairy base until the friendly *acidophilus* have managed to produce sufficient lactase to reverse the reaction to dairy products.

Choosing a Bacterial Supplement

OK, so you feel that you need to start taking some form of bacterial supplement, but what? You go to the refrigerator section of your health-food store and the choice becomes a thing of confusion. The average store employee is of little help simply because they themselves are in the dark.

Considerations for Supplementation

There are a few rules to adhere to if you are serious about supplementing your existing friendly bacteria with strong, viable, living bacteria.

- These are living organisms that should be dormant until you take them; therefore, it is critical that they be stored under refrigeration. If they are not in the fridge, don't buy them.
- Plastic containers are porous and will allow moisture from the fridge to enter and compromise the snoozing bacteria. Your supplement should be in a glass bottle.
- Some bacteria are more aggressive than others. The only two that I have found to be happy cohabiting are L. *acidophilus* and B. *bifidum*. Although the bacteria are snoozing, they could have been left out of the fridge for some time prior to reaching the store. If this is the case, though still sluggish, they will have become active enough for the stronger, more aggressive ones to attack and gobble up weaker bacteria.
- The only bacteria you really need to supplement are L. *acidophilus* and B. *bifidum*. When undertaking serious therapy, use L. *bulgaricus*. Remember that L. *bulgaricus* are strong and aggressive bacteria and should be stored on their own. Any combination that includes these bacteria will eventually contain only these bacteria.
- B. *bifidum* is the dominate bacteria for newborns, nursing mothers, children up to age seven or eight and total vegetarians; therefore, any bacterial supplement for this group must include B. *bifidum* to be complete.

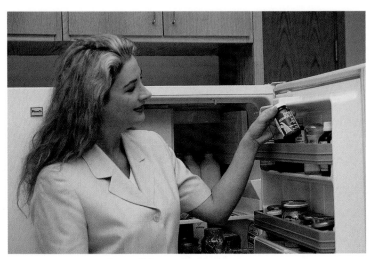

Bacterial supplements should ideally be stored in the refrigerator.

How Important is Refrigeration?

After bacteria are harvested they are freeze dried. This places them in a dormant state prior to bottling, sort of like hibernation. They are suspended within powdered food that may be either dairy based or, for people who are lactose intolerant, non-diary based; often rice powder is used. Although they are dormant, the bacteria are still living, and so need some food to survive. By keeping the bacteria under refrigeration their consumption of this food is minimal; in these conditions it can sustain them in a viable state for up to two years.

When bacteria are not shipped and stored under refrigeration, they are no longer hampered by low temperatures. They become active and readily consume the powdered food that surrounds them. Once this is gone their instinct for survival takes over and they attack their neighbors, the stronger ones destroying and consuming the weaker ones. It isn't long before even the strong ones, bottled up without any other food source, themselves perish. Bacterial supplements claiming that no refrigeration is necessary have a very short shelf life.

Therefore it is ideal to have the supplements refrigerated at all times. However, a short period of time without refrigeration does not ruin the effectiveness of the product, it simply makes its shelf life shorter.

Lifestyle Choices: Regaining Control . . . .

To determine what we can do to maintain or regain our good health, we must first take a good look in the mirror. Prevention is always the best course of action, but it might be too late for that in some cases. I think it is safe to assume that many of us, regardless of age and apparent state of health, need to change our dietary habits to include foods that will support our community of friendly bacteria, add a bacterial supplement to our existing diet, or run as fast as we can to a knowledgeable health professional for guidance in both areas.

A Healthy Diet Means Healthy Bacteria Levels

Having a healthful, balanced, whole-food diet does not happen overnight. It starts at a very early age: good eating habits learned in childhood will carry through for the rest of our lives. The same holds true for poor dietary habits learned in childhood. We often hear people say, "I eat the same way my parents and grand-

Good eating habits learned as a child will ensure a balanced, whole-food diet and a healthy balance of friendly bacteria in the long term.

parents did; they were healthy as horses, so what was good for them is certainly good for me." But food today is not always the same as it was years ago. "You are what you eat" has never been more true. After growing vegetables in the same soil for decades without compart added (centuries in some cases)the soil is simply worn out and no longer carries enough nutrients to give to plants that grow there.

Many of the essential nutrients that our friendly bacteria look for to convert into usable materials for us simply are no longer present. In an attempt to increase yields, many farmers depend on chemical fertilizers (not organic farmers, of course).

The whitish film on organic fruit and vegetables is good bacteria that promotes health.

The drawback to these chemical fertilizers is their effect on the soil nutrients, in particular essential minerals. Chemical fertilizers, when added to the soil, form bonds with the minerals in the soil. This prevents the minerals from being absorbed into the plants growing in this soil. Vegetables grown in these conditions may look great. Pesticide use means that they won't have an insect mark on them but something else they do not have are significant levels of the essential minerals we need to ensure a healthy diet.

Chemistry kills micro-organisms which open the soil and make nutrients available to the plants. The very nutrients that support our friendly internal micro-flora often are either at insignificant levels or completely absent in mass-farmed foods.

The best chance we have to ensure a healthful diet is to shop in stores that guarantee that only organic foods are carried. While this is the ideal, it is not always practical. We may not be able to ensure that every food we eat can be classified as "healthy," but as long as we concentrate on good, nutritious foods as the main part of our diet and avoid foods that are detrimental to good health, we are on the right track. After all, we are all creatures of habit and sometimes the foods that are bad for us are the ones we really hate to give up. But these foods are usually what feed the bad guys in our bodies, particularly the yeasts. To work toward a good, healthy diet, and to support our friendly bacteria, we must make a sincere effort to limit the no-nos.

What Goes In Must Come Out

In addition to helping children form healthy dietary habits, we need to instill in them an awareness of, and practical habits of, regular elimination (bowel movements). When a child says "I gotta GO," take the child to a washroom immediately and start the first phase of their training toward normal regularity. Don't fall into the habit of so many people, who tell their child "we can't go HERE! Hold it until we get home." And often, by the time they get home, the waste material that the child's body was trying to get rid of has been pulled back up into the storage pocket (sigmoid colon), away from the exit. All that has been accomplished is that the sigmoid colon has simply stretched to accommodate the waste. Each time we delay the normal expulsion of waste, the sigmoid gets a little bigger, making the subsequent movement (when it finally comes) much larger than it should normally be. Ever wonder why some children develop fissures and damaged tissue in their rectal area?

Foods That Feed the Bad Guys

The foods that feed the toxic bacteria and yeasts within our bodies are ones that every nutritionist will tell you to avoid. Animal fats, deep-fried fast foods, hydrogenated oils (most commercial vegetable oils), processed meats, rancid oils and nuts, red meats, simple carbohydrates, sugars, doughy breads, pastries and many of the processed dairy foods. This may sound rather ominous, but if you wish to maximize your chances of achieving optimum health, these food categories should be avoided. Since many of these processed and altered foods contain trans fatty acids, they throw an added burden on our already overworked friendly bacteria and cleansing organs.

Deep-fried fast foods feed the toxic bacteria and yeasts within our bodies.

These food groups hurt us in several ways: they increase our level of bad bacteria, they can deposit potentially cancer-causing substances in our bodies, and they don't support our friendly bacteria. When our good bacteria, immune system and cleansing organs are under stress, the bad guys within our internal garden flourish and multiply. You have a choice: avoid foods that contribute to internal stress, or feed the bad guys. You really can't do both.

How About Yogurt?

Isn't yogurt a good source of friendly bacteria? Yogurt is an excellent food, no question. The milk is predigested by friendly bacteria, which makes it easier for our bodies to utilize the nutrients. Unfortunately, most popular commercial yogurts undergo pasteurization after they are manufactured, thereby rendering the bacteria deceased.

If you make your own yogurt, you at least have live bacteria in the finished product, usually L. *bulgaricus* or S. *thermophilus*. But the problem here is similar to the trouble with commercial yogurts. In homemade yogurt, the bacteria, although alive, are simply worn out and have little therapeutic value. Yogurt is a food and is treated as such by our bodies. When food is on the way to the stomach, a signal is sent that alerts the acid-producing glands within the stomach. Stomach acids are produced to break down the food into usable material before sending it on its way down the intestines. But during this process, most of the worn-out bacteria in the yogurt are destroyed.

Natural yogurt as a food source is excellent, and as a carrier for powdered bacteria for inserts for the vagina and colon it is without rival (never use yogurt with gelatine or guar gum added). But as a source of friendly bacterial implantation when eaten, yogurt is not the best choice.

Foods That Support the Good Guys

Fresh, raw vegetables are always at the top of the list of bacteria-friendly food because they already contain living, friendly bacteria looking for a home. Nature put it there for us to consume. Have you ever wondered what the whitish film on some home-grown or organic fruits and vegetables (especially cabbage, grapes and apples) is? It's bacteria—don't wash it off.

Vegetables such as cabbage, broccoli, kale and leafy greens promote and manufacture lactic acids, which our friendly bacteria thrive on. Cold-pressed unrefined oils on our salads, fresh nuts, fruits and whole grains round out the list. Yogurt, while not the best source of viable living bacteria, is still an excellent food and is very supportive to our existing friendly bacteria.

It can be difficult these days to get enough fresh, organic bacteria-friendly foods. It was easier for our ancestors: without refrigeration, they ate more fermented foods, and these contain

Fresh, raw vegetables are always at the top of the list of bacteria-friendly food because they already contain living friendly bacteria looking for a home.

large numbers of friendly bacteria. And of course, all their food was organically grown. This is why we need supplements today.

It's the lactic-acid bacteria on cabbage (the whitish film) that starts the sauerkraut fermentation process.

Recipes To Support Friendly Bacteria ......

We've compiled a list of recipes that will help you attain your goals, starting with a bacteria-boosting juice, the "Rejuvenator."

When choosing recipes that support and strengthen our community of friendly bacterias, we must also look to recipes that will contribute to our overall good health.

Hopefully you will agree that the recipes that I have chosen are not only good for you but are good tasting as well. To round out our diets I have tried to include recipes that will fill complete meal requirements. It is important to be aware that some of the recipes have ingredients that are effective blood and tissue cleansers, therefore it becomes very important to ensure you maintain daily eliminations to keep the toxins that will be drawn out of your body and deposited in the colon from building up. Regularity is a key factor in supporting your friendly bacterial community within the colon. Remember, a healthy colon means a healthy body.

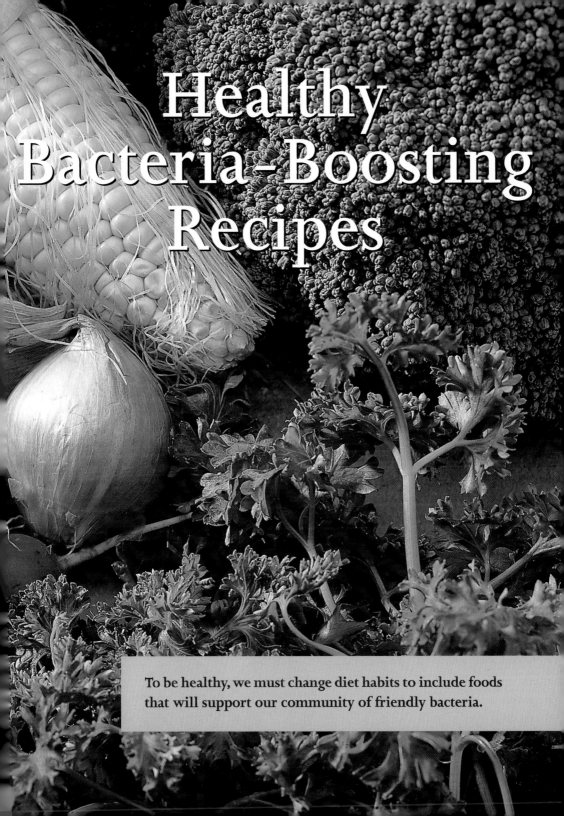

Healthy Bacteria–Boosting Recipes

To be healthy, we must change diet habits to include foods that will support our community of friendly bacteria.

The Rejuvenator

Jump-start your friendly bacteria with this wonderful recipe. I call it "liquid gold," because it is so valuable for rejuvenating your supply of friendly bacteria.

2 cups (500 ml) filtered water

3 ½ cups (875 ml) cabbage, chopped

2 cloves fresh garlic, crushed

Pinch sea salt

1 tsp honey (optional)

Put all the ingredients in a blender. Blend at low speed, gradually increasing to high speed for 30 seconds total. Put the mixture into a covered jar, let stand at room temperature for 3 days. After 3 days, strain solids off. What you have now is your "liquid gold" rejuvenator.

Set aside one quarter cup of the liquid for a subsequent mix. The balance should be drunk over a one day period with meals. Although the mix tastes pleasant (similar to whey or carbonated water) you may wish to add a little honey.

While the initial mixture takes 3 days to mature, subsequent mixes using a portion of the previous liquid takes only 24 hours. Simply follow the original recipe, reducing the initial water by one-quarter cup, blend for thirty seconds, pour into a jar, then add the one-quarter cup of liquid from the previous mixture. Cover, shake well and let stand overnight. It is now ready to strain and use. Remember to set aside the one-quarter cup from this mix also so you can continue to make The Rejuvenator a part of your daily routine.

Serves 1

garlic

red cabbage

Red Grape Cereal and Muffins

Enjoy this breakfast knowing it's very friendly to your body and will give you lots of energy to start the day.

½ lb (250 g) **red grapes**

1 cup (250 ml) **natural organic yogurt or kefir**

1 cup (250 ml) **whole wheat cereal**

Muffins:

½ cup (125 ml) **whole flax seeds**

1½ cups (375 ml) **natural buttermilk or kefir**

2 cups (500 ml) **whole wheat flour**

½ cup (125 ml) **flax meal**

½ cup (125 ml) **natural sugar crystals** (Sucanut or Rapadura)

1 tsp **baking powder**

2 tsp **baking soda**

½ tsp **sea salt**

2 **free-range eggs**

¼ cup (60 ml) **almond oil or melted butter**

1 tsp **pure vanilla extract**

To prepare the muffins, soak whole flax seeds in buttermilk for 2 hours. Preheat oven to 375°F (190°C).

In a large bowl, combine flour, flax meal, sugar, baking powder, baking soda and salt. In a separate bowl, beat eggs and stir in almond oil, vanilla and the buttermilk-flax seed mixture.

Add the liquid ingredients to the dry ingredients and gently stir until smooth.

Pour the batter into 16 medium muffin cups so that they are no more than three-quarters full, and bake for 20 minutes. Let sit on a cooling rack for at least 5 minutes before serving.

Combine the cereal, grapes and yogurt in bowls and serve with the muffins.

Serves 2; yields 16 medium-size muffins

Variation: Substitute the buttermilk in this recipe with kefir—another cultured milk product.

red grapes

You can easily make your own buttermilk. Pour 2 tablespoons of cider vinegar into a measuring cup and add enough organic milk to make 1 cup. Let it sit for 10 minutes and you'll have your buttermilk.

Spring Greens Soup

This wonderfully fresh soup is sure to become a favorite. Served as an appetizer it stimulates digestive enzymes and promotes healthy bacteria. Served as a meal it is both tasty and satisfying.

1 lb (500 g) mixed greens (escarole, chicory, spinach, romaine and watercress), **cut in strips**

¼ cup (60 ml) organic butter

12 green onions, chopped

2 cloves garlic, crushed

2 medium new potatoes, peeled and chopped

2 small carrots, thinly sliced diagonally

1 cup (250 ml) **carrots, diced**

1 cup (250 ml) **green beans, diced**

1 cup (250 ml) **Brussels sprouts, quartered**

1 cup (250 ml) **celery, diced**

6 cups (1.5 l) **vegetable stock**

½ cup (125 ml) **natural cream**

¼ cup (60 ml) **parsley, minced**

¼ cup (60 ml) **lovage**

¼ cup (60 ml) **Parmesan cheese, grated**

In a large pot, heat butter over medium heat and sauté onions and garlic until tender. Add remaining vegetables and sauté for 5 minutes then add vegetable stock, salt and pepper. Bring to a boil, cover and simmer for another 5 minutes or until vegetables are tender. Stir in cream and mixed greens and heat for an additional 5 minutes but do not boil. Garnish with parsley and cheese.

Serves 6

green onion

parsley

Russian Beet Soup

4 medium beets, grated

2 tbsp cold-pressed olive oil

1 large onion, chopped

2 cloves garlic, minced

2 cups (500 ml) white cabbage, shredded

1 cup (250 ml) carrots, diced

2 cups (500 ml) beet liquid

4 cups (1 L) organic vegetable stock

Sea salt and freshly ground pepper, to taste

2 tbsp apple cider vinegar

½ cup (125 ml) celery, finely chopped

½ cup (118 ml) natural sour cream

2 sprigs fresh dill, minced

In a large pot, heat oil over medium heat and sauté onion and garlic until tender. Add beets, cabbage, carrots, beet liquid and vegetable stock; season with salt and pepper. Cover and simmer for 20 minutes.

Remove the soup from heat, stir in vinegar and chill in refrigerator for several hours. Before serving, stir in celery and season.

Serves 6

Zucchini Focaccio

1 cup (250 ml) onion, chopped + 2 tbsp grated

1 tbsp + 1 tbsp cold-pressed olive oil

3 cups (750 ml) whole wheat flour

1 cup (250 ml) zucchini, shredded + 1 cup (250 ml) sliced

⅓ cup (80 ml) natural sugar crystals

5 tsp baking powder

½ tsp baking soda

1½ tsp sea salt

½ cup (80 ml) organic butter, melted

1 cup (250 ml) buttermilk or kefir

2 large free-range eggs

Preheat oven to 350°F (180°C). Lightly grease a 12" (30 cm) pie plate with 1 tablespoon olive oil and set aside. Heat 1 tablespoon olive oil in a pan and sauté chopped onion until tender. Set aside.

In a large bowl, mix flour, shredded zucchini, sugar, baking powder, soda and salt. In a separate bowl, combine butter, buttermilk, eggs and grated onion and stir until smooth. Pour the wet ingredients into the dry and stir until just mixed (mixture may appear dry). Spread the dough in the pie plate and place sautéed onion and sliced zucchini over top. Bake for 55 to 60 minutes, or until focaccio pulls away from the sides of the pan.

German Cabbage Salad

This tasty traditional salad contains a significant amount of lactic acids, which our friendly bacteria thrive on.

12 cups (3 L) **red cabbage, finely shredded**

⅔ cup (165 ml) **filtered water**

⅔ cup (165 ml) **cider vinegar**

½ cup (125 ml) **cold-pressed flax seed oil**

2 tbsp natural sugar crystals

2 tsp caraway seeds

½ tsp Herbamare, original or spicy

Cover the cabbage with boiling water and let stand for about 5 minutes; drain and set aside.

In a large bowl, combine the remaining ingredients and mix with the cabbage. Let stand for 1 hour or longer. (For the ultimate in flavor, make this salad a day in advance, letting it sit in the refrigerator.) Serve cold but not chilled.

Serves 12

red cabbage

Rapadura

Natural Sugar

Natural sugar crystals may be equally substituted for the white sugar called in your recipes. There are many types of natural sugar crystals on the market. Some are superior to others simply because of the way they're made. I use either Sucanat or Rapadura, which is dried cane juice and totally unrefined. Unlike the process used to make white refined sugar, the process used to make these natural sugars preserves the natural rich flavor and nutrition, without preservatives or additives, and actually results in a lower level of sucrose than refined sugar.

Vegetable Delight

Fresh vegetables contain living, friendly bacteria. Vegetables also supply a good amount of fiber, which helps to keep the colon clean. Health benefits aside, this vegetable dish is truly delightful.

1 cup (250 ml) **baby lima beans**

1 tbsp + 2 tbsp organic butter

½ cup (125 ml) **white onion, chopped**

2 cloves garlic, minced

2 carrots, peeled and chopped

½ cup (125 ml) **green beans, chopped**

1 small yellow squash, peeled and sliced

4 stems green onions, chopped

Herbamare to taste

Soak lima beans in lukewarm water overnight or at least 2 hours. Drain and place in a pot of fresh water and cook for 30 minutes. Remove from heat, drain and set aside.

In a large saucepan, heat butter and sauté onion and garlic until tender. Add carrots, green beans, lima beans and a pinch of salt then pour just enough boiling water to cover the vegetables. Simmer covered for about 10 minutes then add squash and green onions. Cover and simmer for another 10 minutes or until the vegetables are tender. Drain, add butter, salt and pepper to taste and enjoy.

Serves 4

string green bean

white onion

Reserve the vegetable water and add it to recipes calling for vegetable stock.

Jerusalem Artichokes with Tomatoes

The Jerusalem artichoke is a tuber that looks like a small, knobby potato or celery root. It has a very elusive flavor and is an excellent supporter of friendly bacteria. Along with the healthy tomato, there's no going wrong with this tasty and healthful dish.

2 cups (500 ml) Jerusalem artichokes

Pinch sea salt

2 tbsp organic butter

2 tbsp cold-pressed olive oil

½ cup (125 ml) white onion, chopped

2 cloves garlic, minced

1 cup (250 ml) tomatoes, chopped

Herbamare to taste

1½ tsp fresh basil, minced

Cook the Jerusalem artichoke in salted boiling water for 10 to 15 minutes or until tender. Drain, rub off skins, slice and set aside.

In a large saucepan, heat butter and oil over medium heat and sauté onion and garlic until tender. Add tomatoes, salt and pepper to taste, and basil. Cook over low heat, stirring continually, until tomatoes blend into the sauce. Add artichoke slices and reheat only until hot.

Serves 4

tomato

Herbamare

Sauerkraut-Cocoa Sherbert

Don't tell your family or friends there's sauerkraut in this luscious dessert until they've finished eating it. They'll never guess there is. Their taste buds will enjoy the treat, while their bodies benefit from the friendly bacteria that's promoted.

4 cups (1 L) **sauerkraut**

¼ **cup** (60 ml) **whipping cream** + ¼ **cup** (60 ml) **whipping cream, for garnish**

¼ **cup** (60 ml) **water**

4 tbsp natural sugar crystals

1½ **tbsp cocoa powder**

1 tsp vanilla extract

¼ **cup** (60 ml) **unsweetened chocolate chips**

2 egg whites, beaten until fluffy and foamy

Fresh mint leaves, for garnish

Strain sauerkraut then put it through a juicer. Set aside.

In a saucepan, heat cream, water, sugar and cocoa on very low heat, stirring continually, for 5 minutes or until just warmed. Remove from heat and place in a bowl of ice to cool then add vanilla extract, sauerkraut juice; mix thoroughly. Add chocolate chips, mix again, then gently fold in egg white.

Pour the mixture into glasses and place in the freezer for 4 to 5 minutes until stiff. Remove from the freezer, garnish with whipping cream and mint leaves and serve.

Serves 4

references

Buchanan, R.E., and N.E. Gibbons. *Bergey's Manual of Determinative Bacteriology*, 8th ed., Williams & Wilkins, 1993

Chaitow, Leon, and Natasha Trenev. *Probiotics: How Live Yogurt and Other 'Friendly Bacteria' Can Restore Health and Vitality*. London: Thorsons, 1990.

Crooke, William G. *The Yeast Connection*. Jackson, TN: Vintage Books, 1998.

Granger, D.N., and J.A. Barrowman. *Gastrointestinal Physiology*. Mosby-Year Book, Incorporated, 1996.

Metchnikoff, E. *The Prolongation of Life*. New York: G.P. Putnam & Sons, 1908.

Moore, D. *Yeast Infections: Nuisance or Nightmare?*

Queen, H.L. *Restoring the Natural Balance of Gastrointestinal Microorganisms.*

Sehnert, K.W. *The Garden Within*, HealthWorld, Inc., 1994.

Shahani, Khem. *Facts and Fallacies of Probiotics.*

Trenev, Natasha. *Probiotics: Nature's Intestinal Healers*. Garden City Park, NY: Avery, 1998.

sources

Department of Food Science and Technology
University of Nebraska
Lincoln, NE
68583 USA
Tel: (402) 472-2815
Fax: (402) 472- 1693

for kefir maker:
Teldon of Canada Ltd.
7432 Fraser Park Drive
Burnaby, BC
V5J 5B9
Order Line: 1-800-663-2212
E-mail: teldon@ultranet.ca

for DDS-1:
Innovite
97 Saramia Crescent
Concord, ON L4K 4P7
Tel: (905) 761-5121
Fax: (905) 761-1453
(or ask at your health food store)

Natren
310 S Willow Ln
West Lake Village, CA
91361 USA
Tel: (805) 371-4737 Ext. 110
Fax: (805) 371-4742

UAS Laborities
5610 Rowland Rd #110
Minnetonka, MN
55343 USA
Tel: (612) 935-1707
Fax: (612) 935-1650
Dr. S.K. Dash, President

First published in 2000 by
alive books
7436 Fraser Park Drive
Burnaby BC V5J 5B9
(604) 435–1919
1-800–661–0303

© 2000 by *alive books*

Book Design:
 Liza Novecoski
Artwork:
 Terence Yeung
 Raymond Cheung
Food Styling/Recipe Development:
 Fred Edrissi
Photography:
 Edmond Fong (recipe photos)
 Siegfried Gursche
Photo Editing:
 Sabine Edrissi-Bredenbrock
Editing:
 Sandra Tonn
 Donna Dawson

Canadian Cataloguing in Publication Data

Moore, Dalton
 Friendly Bacteria

(alive natural health guides, 18
ISSN 1490-6503)
ISBN 1-55312-016-7

Printed in Canada

Revolutionary Health Books

alive Natural Health Guide

Each 64-page book focuses on a single subject, is written in easy-to-understan language and is lavishly illustrated with full color photographs.

New titles will be published every month in each of the four series.

Self Help Guides

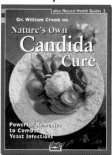

Healthy Recipes

Healing Foods & Herbs

Lifestyle & Alternative Treatments

other titles to follow:

- **Nature's Own Candida Cure**
- **Natural Treatment for Chronic Fatigue Syndrome**
- **Fibromyalgia Be Gone!**
- **Heart Disease: Save Your Heart Naturally**

other titles to follow:

- **Baking with the Bread Machine**
- **Baking Bread: Delicious, Quick and Easy**
- **Healthy Breakfasts**
- **Desserts**
- **Smoothies and Other Healthy Drinks**

other titles to follow:

- **Calendula: The Healthy Skin Helper**
- **Ginkgo Biloba: The Good Memory Herb**
- **Rhubarb and the Heart**
- **Saw Palmetto: The Key to Prostate Health**
- **St. John's Wort: Sunshine for Your Soul**

other titles to follo

- **Maintain Health v Acupuncture**
- **The Complete Natural Cosmetic Book**
- **Kneipp Hydrother at Home**
- **Magnetic Therapy and Natural Heal**
- **Sauna: Your Way Better Health**